ISBN 978-1-332-04642-3
PIBN 10275397

# 1 MONTH OF
# FREE
# READING

at

## www.ForgottenBooks.com

By purchasing this book you are eligible for one month membership to ForgottenBooks.com, giving you unlimited access to our entire collection of over 700,000 titles via our web site and mobile apps.

To claim your free month visit:
www.forgottenbooks.com/free275397

English
Français
Deutsche
Italiano
Español
Português

# www.forgottenbooks.com

**Mythology** Photography **Fiction**
Fishing Christianity **Art** Cooking
Essays Buddhism Freemasonry
Medicine **Biology** Music **Ancient**
**Egypt** Evolution Carpentry Physics
Dance Geology **Mathematics** Fitness
Shakespeare **Folklore** Yoga Marketing
**Confidence** Immortality Biographies
Poetry **Psychology** Witchcraft
Electronics Chemistry History **Law**
Accounting **Philosophy** Anthropology
Alchemy Drama Quantum Mechanics
Atheism Sexual Health **Ancient History**
**Entrepreneurship** Languages Sport
Paleontology Needlework Islam
**Metaphysics** Investment Archaeology
Parenting Statistics Criminology
**Motivational**

# A CONTRIBUTION TO THE LIFE-HISTORY OF LILIUM TENUIFOLIUM ♭

BY

## DANIEL DA CRUZ

_____.

_._

*THESIS SUBMITTED TO THE FACULTY OF SCIENCES OF THE CATHOLIC ,*
*UNIVERSITY OF AMERICA, IN PARTIAL FULFILLMENT OF*
*THE REQUIREMENT FOR THE DEGREE OF*
*. DOCTOR OF PHILOSOPHY*

_____

WASHINGTON

1915

PRESS OF
H. L. & J. B. McQUEEN, INC.
WASHINGTON, D. C.

# TABLE OF CONTENTS.

# A Contribution to the Life-History of Lilium tenuifolium.[1]

## INTRODUCTION.

Among the cytological problems so far investigated none has awakened the interest of the investigator to such a degree as those dealing with the karyokinesis of maturation. If the true interpretation of hereditary phenomena is concerned, as it appears to be, with the reduction division, such interest in this field of investigation is sufficiently justified. To throw some further light upon the subject by bringing out the facts observed in a new and exceptionally favorable object is the main reason for the present investigation.

The most important period in connection with maturation in Lilium is that which embraces the whole prophase up to the anaphase or telophase of the first mitosis. It has been the subject of the minutest investigations of almost all cytologists, and concerning it the most different hypotheses have been constructed. One of the chief subjects of the present work also will be mainly those phenomena leading to the reduction division of the chromosomes of the first mitosis. The further development, however, of the microsporocyte and megasporocyte up to the conjugation of the oosphere and sperm will not be neglected but occasionally considered and spoken of by way of illustration and completion.

The plant finally selected for the present study was *Lilium tenuifolium*, which proved to be exceptionally well adapted to giving the most remarkable results. The nuclei of the germ-cells are large and all their nuclear contents very conspicuous and well defined. The gathering of the flowers began in the first days of May and lasted to the beginning of June. Occasionally new collections were made in March and April of the following year from new sets of lilies grown in the green-house, which proved to be as good as those grown in the open air.

---

[1] Contribution from the Biological Laboratory of the Catholic University of America, No. 1.

The flowers were gathered in all stages of their development from the time their anthers were but one millimetre in length, up to five days or more after pollination. In the youngest material the microsporocytes were already in synapsis or nearly so. Then the successive stages of the microsporocyte followed one another so rapidly that within two weeks after that time, that is, when the flower was but ' half grown, the pollen grain was already completely formed. The hour in which to gather the flowers seems to be of no consequence as the whole development of the microsporocyte goes on gradually and requires from one to two weeks for its completion.

For the megasporocyte the situation is different. The stages previous to synapsis occupy the time until about the opening of the flower. Then all the stages up to the second division follow very closely so that in about two or three hours they are complete. Thus in one single ovary gathered at sunrise, when the flower was opening, all the phases between synapsis and the 4-nucleate stage of the embryosac were found. The later stages also followed very rapidly. Fecundation took place ordinarily between three and five days after pollination, according to temperature conditions.

Anthers and pistils were carefully separated from the flowers, dipped in 95 % alcohol and instantly submerged in Schaffner's killing fluid. After the tissue was killed and thoroughly washed, it was passed by degrees through alcohol of 10 %, 20 %, 30 %, etc., up to absolute alcohol. Imbedding in paraffine was made through chloroform in the usual way. The sections of the microsporocyte were made not less than 12 $\mu$ thick. Some, however, were cut 6, 10, and 16 $\mu$, for comparison. The later stages of the embryosac were ordinarily cut 18 and 20 $\mu$; the younger stages 12 and 15 $\mu$.

The stains used were Delafield's haematoxylin, Fleming's triple stain, and sometimes iron-alum-haematoxylin. Delafield's haematoxylin proved to be the best stain for general purposes, but Heidenhain's iron-alum-haematoxylin was of special value for chromatic contents, especially for the nucleoli.

All the figures were drawn with the camera lucida, at table distance, with a Bausch and Lomb ocular number 10 and oil immersion objective 1.9 mm., and Leitz compensating oculars numbers 8 or 12 and apochromatic oil immersion objective number 2 mm. The figure in synapsis observed in living tissue was drawn with the camera lucida with a Bausch and Lomb ocular number 15 and a 4 mm. dry objective. Other powers used will be marked after the respective figures.

I am especially indebted to Professor J. B. Parker of the Catholic University of America for the material which he put at my disposal and for the many suggestions imparted to me in the course of the work, and also to Mr. H. H. Bartlett of the Bureau of Plant Industry at Washington, D. C., for many helpful criticisms.

# INVESTIGATION.

## I. THE MICROSPOROCYTE.

A. From resting stage to synapsis.—Towards the end of the last sporophytic division the microsporocytes resulting from it are closely united to one another in the central cavity of the anther, surrounded by two rows of tapetal tissue. In that stage the microsporocytes do not differ essentially from the neighboring somatic tissue, except for their more regular shape and deeper stain. The appearance of both cytoplasm and resting nucleus is more uniformly reticulated and more dense than is the case in the neighboring tissue.

The beginning of the prophase (Fig. 1) is characterized by the increase in size of the whole cell, the meshes of the chromatin network becoming wider and all of its structures showing an aspect more characteristic and definite. As a whole the nucleus presents the form of a large knot very rich in chromatic substance floating in the nuclear enchylema, containing three to five and sometimes more, large, deeply stained nucleoli, suspended in the meshes of the reticulum. In the progress of its developments (Fig. 2) the chromatic network begins to loosen from its many anastomoses and to arrange itself into a definite spirem. The threads are very delicate, single in appearance, crossing in all directions, with a single row of prominent chromatic granules without very definite shape and irregularly distributed throughout.

As the prophasic movement goes on (Fig. 3) the fine spirem becomes more conspicuous and definite, more free from its anastomoses, and shows a slight tendency to loosen from the nuclear wall as a remote preparation for the so-called synaptic stage. Up to the stage represented in figure 4 the growth of the nucleus is proportionately greater than that of the general cytoplasm. The chromatic substance diffused in the enchylema begins to vanish from the beginning of the prophase, disappearing almost entirely when the synaptic contraction begins. At the same time the chromatic threads become more conspicuous, keeping, however, the same appearance as before, as far as the synaptic stage, except that their chromatic granules remain of the same size or even become smaller than in the previous stages.

All the figures so far mentioned present some special features when stained with Heidenhain's iron-alum-haematoxylin. The nuclear reticulum with its chain of chromatic granules, and the numerous nucleoli, with one to three or more big vacuoles, are the only structures shown in the nuclear cavity (Fig. 4). In the foregoing stages no filaments were seen in pairs or showing any evidence of a beginning of zygosis. Figures 57 to 60 of the megasporocyte, which may be placed between figures 4 and 5 in order of development, seem, however, to contradict what has just been said. Those figures are, however, the ones which gave the best evidence against the hypothesis of a longitudinal conjugation of the spirematic threads. The clear and single fact, as evinced by the figures, may be stated as follows: The spirem, made up of a single chromatic thread, accomplishes, from the beginning of synapsis, a slow and gradual movement of concentration, the final result of which is a close and compact knot as represented in figure 8. In obedience to the physical conditions it is only natural that the parts of the spirem floating freely in the enchylema should be attracted with greater rapidity toward the center, while the parts attached to the nuclear wall should be attracted more slowly, giving rise to the particular dispositions of the threads represented in figures 57 to 60, and even in a more advanced stage represented by figure 5. In figure 6 the spirem appears double in some places, which is true for different micro-sporocytes of the same anther cavity. A little further in the same anther cavity (Fig. 5) the spirem is single. None of these figures are as yet in full synapsis judging by the character of the cells, the aspect of their nuclei, and the surrounding tapetal tissue. According to the one interpretation the double character of the threads would indicate a longitudinal conjugation, but this view seems to be opposed by the fact that the microsporocytes are not yet in synapsis, where such conjugation should take place. Now we have seen that the reticulum effects the movements of concentration towards synapsis in the form of many loops, which are single to the stage represented by figure 5, and that the same loops are those which appear double a little later when very near synapsis. But, according to the above interpretation, that which in the figures constitutes the doubleness of the spirem, would be the two legs of what constitutes a loop. My interpretation is that an actual longitudinal cleavage takes place after the spirematic development, and that this cleavage indicates the longitudinal division that will take place to form the chromosomes of the second mitisis. The phenomenon is not, however, very conspicuous. It was seen but once in the megasporocyte and only a few times in the microsporocyte, in both cases either before or immediately after full synapsis, disappearing entirely after the beginning of the second contraction. The cleavage is never the apparent result of the two legs of the loops coming into contact, but is a splitting of the same loops which are developed from the contracting reticulum.

According to this view, figures 5, 6, 8, and 10 to 13 represent successive stages of synapsis. The first two approach the full stage of contraction represented in figure 8, while figures 10 to 13 follow it closely but represent the spirem as it unrolls more and more until it occupies again the whole nuclear cavity. The thin threads of figures 5 and 6 become thicker in figures 10 to 13. Is this an effect of longitudinal conjugation between them? It is hard to identify the above mentioned doubleness of the spirem with a beginning of zygosis, and I have never seen anything like it in thousands of figures in synapsis. The difference of thickness of the spirem at the beginning and at the end of full synapsis can be explained only by a growth in diameter of the same threads during complete contraction (Fig. 8), be that growth an effect of the shortening of the threads, of the assimilation of chromatic matter from outside, or of both.

Figure 13 offers us a most remarkable feature in the slow and gradual development or growth in diameter of the spirem during the different phases of synapsis, refuting completely the theory of a conjugation. Its nucleus is still in synapsis, but its projecting chromatic threads appear of three different diameters. There are one or two very thick filaments, two very thin ones, and the prevalent number of a medium diameter. Judging by the character of the tapetal tissue, by the state of separation of the microsporocytes and specially by the general features of the spirem in that cell and its neighbors, the nucleus is just coming

out of synapsis. However, there are still some very distinct thin threads such as appear in the stages previous to synapsis, in contrast with others very thick and characteristic of the last phase of the synaptic stage, but the most striking feature of the figure lies in the fact that there is an intermediate size in the diameter of the spirem (Fig. 14, *b*). This figure was drawn with the greatest care in order to represent the actual diameter of the spirem without any exaggeration. Though the two thin filaments (Fig. 14, *c*) are partly placed side by side, this fact does not seem to me to give any evidence that they are in a longitudinal conjugation. Comparing this figure with those which precede and follow it, the only explanation of their approximation is that they represent two loops more or less in contact. In none of the preparations were filaments ever seen partly united longitudinally and partly moving one towards the other to complete such conjugation. The apparent superposition of two or more filaments or loops is nothing else than a mere accident. Their growth in diameter is never made suddenly but by degrees as clearly appears from figures 13–14.

According to Berghs, Gregoire and generally all the parasyndeticians, the strongest proof of a longitudinal conjugation between the threads during the synaptic stage lies in the fact that the said threads appear thin in the beginning of synapsis, and about twice as thick towards the end of it, *without any intermediate stage in thickness*. Even if this were the case I do not see any strong evidence for the conclusion. But it is not the case. Figure 13 is an instance in which all the gradations are found in the thickness of the spirem, and this is not confined to that one case but is shown elsewhere in the same pollen-sac.

Each of the three different size filaments in figures 13 and 14 has a row of single chromatin granules irregularly distributed throughout. Both the thinner threads and their granules are equally deeply stained, while the middle and thicker ones give the appearance of a linin ribbon supporting more deeply stained chromatin granules. Whether this difference in coloration is due to a different chemical composition or a simple optical illusion, I can not state with certainty.

The foregoing features of synapsis seem to point out its main use in the development of the germ cells. The enchylema of the nucleus is, in its early stage, full of chromatic substance diffused throughout. The substance disappears completely as the chromosomes begin to be arranged. What becomes of this substance? It was undoubtedly taken up by the spirem. On that assumption the synaptic movement serves to bring about the concentration of the chromatic material. That the concentration takes place by degrees is shown by the differences of diameter of the spirem (Fig. 14).

Synapsis was observed by the writer in living material. The microsporocytes of each anther were studied in both the living and fixed conditions; that is, each anther was divided into two pieces, one of which was immediately examined under the microscope in living condition, and the other killed and stained in the usual way for comparison, thus bringing together the facts observed in both sets of material. I arrived at the conclusion that synapsis is a normal phenomenon in the early stages of the prophase, and that the reagents, when properly handled, have very little influence on it; for its appearance is essentially the same in both cases, except that the projecting filaments of the spirem can not be seen in living material (Fig. 9).

B. FROM SYNAPSIS TO METAPHASE.—The most characteristic phase of this period is that which may be called the second contraction (second synapsis of Sargant). Around this centers the arrangement and completion of the chromosomes. After synapsis the spirem occupies the whole nuclear cavity as an uninterrupted series of coils and loops but without any definite orientation (Fig. 15). Up to this stage the spirem appears as a single thread, continuous and apparently without endings. The thickness of the spirem is not very uniform, and there is a chain of chromatin granules irregularly distributed along the threads.

This phase does not last long. The loops, at first very irregular in shape, very broad and with but few twists, begin to orient themselves from the center to the periphery (Fig. 16). The loops shorten much in this stage, and their legs become twisted one to five times. The nucleoli seem to fuse more or less, chiefly in the megasporocyte, and assemble in the middle of the nuclear cavity but sometimes towards one side of it (Fig. 17). The spirem at this stage (Fig. 16) does not yet seem divided into chromosomes and the longitudinal splitting may still be seen in some of its loops (Figs. 16–17), but the general structure is that represented in figure 17. In the following stages the longitudinal fissure tends rather to disappear entirely and the chromatin granules to become confused with the chromosomic substance, but appearing again occasionally in more advanced stages of the second contraction. The shortening of the loops and their simultaneous growth in diameter continues progressively from the stage shown in figure 16 as is well illustrated by figures 17–18, until the twelve chromosomes become entirely free in the nuclear cavity (Fig. 19). This condition of the chromosomes was called *diakinesis* by Haecker, to mean the last stage of the prophase in which the definite chromosomes become completely free in the nuclear cavity and ordinarily in more or less contact with the nuclear wall.

The formation of the loops from the spirem through an uninterrupted and complete seriation can be followed from the loose spirem and even from before synapsis up to the diakinetic stage. In the first place it is evident from figures 16 to 21 that the diakinetic chromosomes come by a gradual succession from the loops of figure 16. There is a perfect seriation through all those figures. That figure 16 comes directly from figure 15 by a simple arrangement and orientation of the parts of its spirem is easy to understand. In the same way a similar gradation is found in the first stages of the prophasic movement (Figs. 4 to 8), especially if compared with the parallel stages of the megasporocyte (Figs. 57 to 60). The only difficulty will be with regard to what happens during full synapsis, but here also we find a complete seriation in the phases shown by the spirem in its behaviour towards the formation and completion of the chromosomes. The synaptic phase covers three main stages: one of contraction, one of rest, and the third of expansion of the spirem. Practically the same number of twists of the spirem appear in both extremes of the synapsis; the median fissure of the spirem, when it occasionally appears, comes along the twists of it both before and after synapsis, without the least evidence of a conjugation between two filaments, but as the temporary splitting of a single thread. The double diameter of the post-synaptic spirem is caused by its shortening, during its resting phase and possibly by some kind of nutrition during that time, as shown in the different sizes in diameter of the same spirem in figure 13.

When free in the nuclear cavity the chromosomes are smooth, homogeneous in structure, without conspicuous chromatin granules, more or less attached to the nuclear wall and not effecting any definite orientation. The structure of the diakinetic chromosomes (Fig. 19) is absolutely unmistakable. They never appear as two superimposed branches except in very few cases, and these are the result of the cutting process. A horseshoe or loop more or less twisted into two or more coils, such is the shape and structure of the diakinetic chromosomes. During this stage the transformation of all the structures of the microsporocyte is very rapid. The chromosomes shorten more and more, become very large in diameter, and orient themselves around the center with their heads turned toward the periphery (Fig. 20). Meanwhile the nuclear wall breaks down and disappears, the general cytoplasm coming into contact with the chromosomes. The nucleoli break down, too, into many micronucleoli and disappear apparently from sight (Fig. 20) or pass to the general cytoplasm (Fig. 69). The behaviour of the nucleoli at this stage therefore seems different in the two sporocytes. But this most probably is only apparent. In the megasporocyte they pass to, and wander freely in, the cytoplasm. In the microsporocyte they are not visible after the rupture of the nuclear wall but this is certainly due to their minuteness as compared with those of the megasporocyte.

*C.* FROM METAPHASE TO DAUGHTER STAR STAGE.—After the twelve chromosomes separate and become free in the nuclear cavity they continue to shorten and grow in diameter, placing themselves radially in the equatorial field to form the mother star (Figs. 20 and 21). Spindle fibres are to be seen in both figures 20 and 21. In the former they appear as wheel rays between each two adjacent chromosomes and are as large as the chromosomes themselves or even larger. In the latter they run from pole to pole as slender filaments ordinarily thicker when they attach themselves to the tops of the chromosomes. The first are certainly the vestiges of the so-called *central spindle*,[1] the *mantle fibres* being visible only in longitudinal sections of the

Wilson. "The Cell in Development and Inheritance" (1911).

microsporocyte. It appears from figure 20 that the mantle fibres are not made up of single and slender fibres, but of bands much like the septa of a sea-anemone. This is true at least for the fibres interposed between any two chromosomes.

The manner in which the chromosomes place themselves in the equator of the figure (Fig. 21), with the extremities of their legs attached to the spindle fibres, and the loop turned towards the periphery, clearly shows how the division of the bivalent chromosomes, the true reduction division, will take place. The chromosomes represented in figures 22 to 25 give different stages of their development towards division. Following a number of figures of that stage and carefully studying their individual chromosomes we find them at first with their legs more or less closely twisted (Fig. 22) and attached by their extremities to the spindle fibres. In a more advanced stage the spindle fibres begin to contract, pulling and stretching the legs of the chromosomes and drawing them towards the poles, while the chromosomes themselves are becoming more and more untwisted until they assume the appearance shown in figures 29 and 71. While these movements take place the spindle fibres begin to move from the top of each half bivalent chromosome to its center, giving to each half chromosome the U-shaped form which will fully appear in the following stages (Fig. 26).

For the parasyndeticians the end to end union between the two branches of the bivalent chromosomes is merely apparent, as a result of a close approximation of their two extremities. But this can not be thought of in our material. The features observed in figures 20 to 25 are not exceptions. They represent the real facts throughout all the material. Two branches more or less parallel, crossed or twisted, but free in their extremities were the real exceptions and are undoubtedly the result of the cutting process. Furthermore, if those features were simply the result of a close approximation of the extremities of any two branches, why should not this approximation and close contact take place toward the center as well as toward the periphery of the star? The loop is, however, and without exception, always turned outside, provided the full metaphase stage has already been reached. On the other hand if we consider the extremely large volume of the chromosomes in this stage, that their looped heads are still bigger and that they are all turned outside, it is to be expected that those heads would be cut off when cut in very thin sections, giving by this means the illusion of two branches more or less parallel and slightly attached to one another. That this is actually the case for the great majority of the figures presented by the parasyndeticians, I am inclined to think, because of the fact that their preparations were ordinarily cut too thin, 8 and 5 μ and even less. My acquaintance with the literature of the European cytologists and my personal acquaintance with at least one disciple of the Louvain school [1] enables me to take that thickness as the average used in their preparations. In this manner the appearance of two branches in the diakinetic chromosomes and even in previous stages is likely to be artificial and illusory. This accident is avoided in my preparation by cutting them to 10, 12, 16, and 18 μ in thickness according to their development both vegetative and ontogenic. In other words, the apparent form of the chromosomes up to metaphase can be produced and is really caused in many instances by the cutting process, while their looped form cannot happen by such accidents.

For our material I may state with the greatest certainty the following facts:

1. That the definite chromosomes form perfect loops like a horseshoe, their branches being more or less twisted with only two free ends (Figs. 19 to 25).

2. That the diakinetic chromosomes are identical with the loops of the preceding phases, although modified by the contraction of the filaments and accompanying growth in thickness.

3. That the separation of the definite bivalent chromosomes in the anaphase to give rise to the daughter stars, is made by a transversal breaking on the middle of the horseshoe-shaped chromosomes (Figs. 22 to 25 and 29).

These considerations lead us to the conclusion that the definite chromosome is made up of two pieces, the somatic chromosomes, fused end to end with one another. They exclude completely the zygosis of the parasyndeticians and establish the truth of the metasyndetic hypothesis, which admits an end to end conjugation of the somatic chromosomes.

[1] Martins Mano.

Thus far it has been impossible to prove between what kind of chromosomes this conjugation takes place or whether it really takes place at all. But in view of the facts just outlined no other hypothesis can be conceived. The first conception of Schaffner, afterwards developed by other cytologists and himself for *Lilium philadelphicum*, *L. longiflorum* and *Erythronium* seems definitely established, at least for *L. tenuifolium*.

After the chromosomes place themselves in the equatorial plane and form the mother star they begin to be pulled apart by the spindle fibres. The insertion of the spindle fibres ordinarily takes place on the extremities of the legs of the chromosomes, seldom towards the center of them. Now, as the spindle fibres contract the chromosomes untwist gradually and become more or less straight at the middle and break apart finally at the middle into two halves. When the chromosomes are stretched in the center their legs take ordinarily the U-shaped form, but in some cases they appear as a more or less straight line as shown in figures 29 and 71. Even in this case each univalent chromosome takes the U-shaped form as soon as they separate from one another. When this stage is reached each univalent chromosome follows its respective course towards the pole with the utmost speed. It is during this time that the spindle ordinarily changes its place of attachment to the loop of the U-shaped or V-shaped chromosomes.

In what manner does this V-shaped form of the daughter chromosomes originate? For Gregoire[1] it originates simply in the longitudinal division of the daughter chromosomes which takes place about the end of the first metaphase or during the anaphase. He states (p. 262) that when the daughter chromosomes are attracted to the poles of the figure by the spindle fibres, attached ordinarily to their extremities, an incomplete longitudinal division takes place from the extremity nearest the equator, that is, opposite to that which is attached to the spindle fibres. This partial longitudinal division separating from below the two halves of the daughter chromosomes gives to each one the V-shaped form. In *Lilium tenuifolium* I was never able to see such longitudinal splitting, either during the metaphase or in the course of the anaphase. Here the splitting of the daughter

---

[1] Gregoire. Les cinèses polliniques chez les lilliacées. La Cellule, 16 (1899).

chromosomes takes place between the second metaphase and the second anaphase, and does not begin in the extremities of the daughter chromosomes but in the middle of them.

It is interesting to note the behaviour of the spindle fibres from the time the chromosomes begin their dividing movement to the beginning of the telophase. At first it seems as if the spindle was an especially organized structure both to support and guide the chromosomes from the mother star to the daughter star stages. The spindle is composed of two parts, one central, the *central spindle*, the other external, the *mantle fibres*. The former is oval; the fibres, running from pole to pole without interruption, seem to act as a kind of skeleton to support the chromosomes, and persist through the complete division of the cell and reconstruction of the daughter nuclei. The mantle fibres are attached to the chromosomes at the tips of the legs and are therefore divided into two halves on the equator of the figure. Those are the fibres which draw the chromosomes apart and drag them to the poles (Fig. 26). They disappear shortly after the daughter star is formed.

In the microsporocyte the nucleoli were lost from sight between figures 20 and 26. This is undoubtedly a result of their excessive minuteness for they appear distinctly in the same stages of the megasporocyte (Figs. 69 and 71). When, however, the reduced chromosomes gather at the poles of the figure (Fig. 27) they again make their appearance within the spindle fibres. (Their behaviour will be fully discussed in a subsequent paper.) At this stage the chromatin granules, which were lost in the previous phase, again make their appearance very distinctly in the great majority of the daughter chromosomes.

Now as the telophasic development goes on (Figs. 28 to 30), the line drawn from pole to pole of the spindle seems to contract towards the equator marking a greater approximation of the daughter nuclei, wherein there is no true reconstitution, the chromosomes maintaining their independence and the chromatin granules becoming more conspicuous (Fig. 30).

Now the separation of the two daughter nuclei progresses and becomes complete by the division of the entire microsporocyte. An equatorial cell-plate begins to be formed, first through the connecting spindle fibres (Fig. 28), afterwards through the whole cytoplasm (Figs. 29 and 30) thus separating the two new cells which become the two daughter cells of the microsporocyte (Fig. 31). The new cell wall thus formed is a secondary one, inasmuch as the two daughter cells remain within the primitive wall of the microsporocyte. During these movements the micronucleoli fuse into a small number of bigger nucleoli and gather together around the nucleus. They probably enter the nuclei at this stage as they do in the case of the megasporocyte but they were not seen to do so.

D. FROM DAUGHTER STAR STAGE TO THE COMPLETION OF THE POLLEN GRAIN.—The succeeding phases follow one another with great rapidity. Figure 31 represents most probably the first movements of the second division of the microsporocyte. The micronucleoli are already expelled from the daughter nuclei and occupy a position very typical and characteristic of this stage. There is no nuclear wall formed and the chromosomes still maintain their individuality and become less closely associated in preparation for the second division. For this purpose the chromosomes form a new nuclear plate equivalent to that of the first mitosis but not so regular and symmetrical. This second division of the chromosomes takes place longitudinally (Fig. 32) as represented in figures 33 to 38 where the gradual process of division is shown in detail. The chromosomes in these figures were taken from longitudinal sections of the microsporocyte and are in perfect accordance with those studied in cross sections of the same. In both longitudinal and cross sections the last phase of the chromosomes before they separate entirely from one another, presents the aspect shown by figures 32 and 39.

Now, attracted by the same means and in the same manner as in the first mitosis, the grand-daughter chromosomes, already separated, begin to move toward the poles sometimes under the form of a U or V but ordinarily as rods very broad and of different sizes (Figs. 39 to 41). This last figure shows in a remarkable manner the rapidity of the phenomena of the last division. On the one hand there are still two chromosomes almost in contact with the equator as if they had just been separated from one another, and on the other hand the other chromosomes have almost joined together into two constituted nuclei, with the micronucleoli gathering around them and a midplate already separating the spindle fibres and giving rise to the tetrad. This stage is better shown in figure 42 where the stage of the nuclei is still more advanced. A new transversal division of the cytoplasm follows immediately and separates the two nuclei, the tetrad being thus formed, each nucleus constituting a new cell with its own cytoplasm and wall (Fig. 43).

The behaviour of the micronucleoli in both divisions of the microsporocyte may be summarized as follows: After the diakinetic chromosomes are formed and the nuclear wall disappears the nucleoli are broken into many micronucleoli and expelled into the cytoplasm. After the complete division of the nucleus takes place the micronucleoli fuse together into a smaller number of larger micronucleoli and gather around and finally within the daughter nuclei where they remain for a short time. But as soon as the movement of the second mitosis begins the micronuclei are again thrown away towards the cytoplasm and again broken into many pieces. When the tetrad begins to separate the same centripetal movement of the micronucleoli with their concomitant fusion is noted, but they do not reach the center of the nucleus until the pollen grain is completely formed (Figs. 44 to 46).

The second mitosis seems to be characterized by special movements of the chromosomes. First they split longitudinally about their center (Figs. 33 and 34) to which the spindle fibres probably attach themselves, pulling the chromosomes apart (Figs. 35 and 36) until they separate at their extremities nearer the equator (Figs. 37 and 38). I was not, however, able to see the spindle fibres attached to such chromosomes. But it is proper to suppose that the spindle fibres will act similarly in both mitoses.

Both in the daughter star and tetrad stages the autonomous chromosomes seem to persist, although more or less anastomosed in the last instance by means of some protoplasmic filaments (Figs. 42 and 43). After a period of apparent rest the common cell wall of the tetrad bursts, liberating the pollen grains, which grow very much in a short time, modifying their shape and protective wall.

Figure 44 represents a grown and complete pollen grain. Its form is oval with a little flatness about the longitudinal opening

(Fig. 46). Its form, the shape and structure of its envelope and its appearance both in cross and longitudinal sections is better shown by figures 44 to 46 than by any kind of description. The chromosomes, doubtless rolled up in all directions, are cut into many pieces but most probably maintain still their individuality although much anastamosed by means of protoplasmic fibres.

## II. THE MEGASPOROCYTE.

The essential phases in the ontogenic evolution of the megasporocyte up to the end of the first mitosis are the same as those observed in the microsporocyte. The main differences between the two sporocytes are those resulting from the bigness of the megasporocyte, which renders more clear and conspicuous the phenomena to be studied. As a whole the facts observed in the megasporocyte are a thorough confirmation of those found and investigated in the microsporocyte. Let us review its ontogenic development in the same order followed in the preceding chapter.

A. FROM THE LAST SPOROPHYTIC DIVISION TO SYNAPSIS.—The megasporocyte proceeds directly from a hypodermal cell of the outermost layer of the periblem of the nucellus. The last somatic division takes place long before the integuments begin to appear. The cell of that division nearest the periphery becomes the megasporocyte without any further division, as is generally the case among the lilies and some plants of other families. After this last division is accomplished the two resulting cells are structurally identical, so as to look like two megasporocytes. It is evident, however, that one of them will remain characteristically somatic, while the other, nearer the center of the nucellus, will become the definite megasporocyte. The only distinction between it and the somatic cells consists in the larger size of the former, its more spherical form and its deeper coloration by haematoxylin.

As soon as the last somatic division is complete the megasporocyte grows exceedingly, affects a more or less oval form with its greater axis perpendicular to the peripheral line of the nucellus, pressing and displacing the surrounding tissue. Its cytoplasm is more deeply stained than that of the somatic cells, with a more voluminous nucleus containing three or more nucleoli. The nuclear reticulum is very fine and extremely entangled in the diffused chromatic substance suspended in the enchylema. The chromatic threads are single, crossing in all

[ 22 ]

directions and apparently without a chain of chromatin granules (Fig. 53).

In the further development of the megasporocyte its nucleus still grows to very large proportions, the threads appear discontinuous here and there and begin to free themselves from the chromatic substance in which they were previously imbedded. Prominent chromatin granules begin to appear throughout the filaments, but their size and distribution is not uniform and regular (Figs. 54 and 55). Figure 56 was stained in iron-alum-haematoxylin. It shows a nucleus still larger than that of the previous figure, the nucleoli are free from the chromatic substance, and the threads, which are showing already a marked tendency to separate from the nuclear wall, are apparently larger in diameter and bear a more regular chain of chromatin granules. All these features are most probably shown because this stain is more likely to show the chromatic elements with accuracy.

Figure 57 represents a megasporocyte at about the same stage of development as the previous one. It is, however, much larger and with more conspicuous minute structures. This is probably due in part to the greater vegetative development of the respective pistil, but the fact that figures like 54 and 55 and others like those represented by figures 62, 64, etc., were found sometimes in the same pistil, points to a different assumption which is based on actual facts. It is my opinion that the synaptic movement begins with figure 57 or 58 and attains its culmination with figure 62. All figures between them are gradual stages of synapsis. All the phases represented in figures 53 to 58 come together at almost the same time in the early stages of the vegetative development of the pistil, whereas the stages represented in figures 68 to 75 may appear all together in the same pistil at the close of its floral development, that is, at about the time the flower is ready to open. It is between figures 57 and 64 that a great space of time intervenes (about three weeks) during which the greater vegetative development of the pistil takes place, although the ontogenic development is very limited. This is the reason why the figures increase so extraordinarily in size from figure 56 on although the evolution of the spirem proceeds slowly and almost imperceptibly.

This circumstance, however, enables us to follow the synaptic movement of the reticulum in its minutest details. The main difference between figures 56 and 57 results from the larger size of the latter, which makes the threads more discontinuous and permits the vacuoles of the nucleoli to be seen more clearly. Figure 58 may be considered as the next stage in the ontogenic development of the megasporocyte. It differs from the preceding one chiefly by the fact that its threads begin to separate more conspicuously from the nuclear wall and move towards the center of the nuclear cavity as the first step towards synapsis. This movement is more accentuated in figure 59. The more the threads get loose from the nuclear wall the greater is their concentration towards one point of the nuclear cavity. Figure 60 represents a more advanced stage of the synaptic movement. Some filaments are still being pulled from the nuclear wall, some others are being strongly carried to the center. Figure 61 is almost entirely in full synapsis. There are no filaments adherent to the nuclear wall, but the characteristic aspect of synapsis is not yet reached. This appears only in figures 62 and 63 where the filaments come in full contact with one another, forming a kind of knot more or less compact.

At the first glance the threads of all these figures seem to be more or less arranged in pairs and placed parallel to one another, as if they were to begin a process of conjugation. But by a closer examination we can see with the greatest clearness that such is not the case. The fact is, of course, that any two parallel filaments are the legs of one loop, and more commonly of two different loops, the other two legs of which cross with those of their neighbors, their outer curve being yet in contact with the nuclear wall or nearly so. It is interesting to note again the special movements of the threads on their way to the center of the nuclear cavity. While they are separating themselves from the nuclear wall they move straight to the center as open loops (Figs. 59 and 60), but as soon as they get loose they roll themselves and twist two or three times (Figs. 60 and 61), as a result of more freedom of movement.

Whether these loops are identical with those which will appear after synapsis, and generally throughout the whole prophase, it is impossible to determine. The important fact to note in this connection is the tendency of the threads to be thrown into loops and coils long before synapsis. But it is immediately before full synapsis (Figs. 60 and 61) that the loops appear in their characteristic shape. On the other hand no true endings are to be seen in the spirem, no parallel threads which cannot be ascribed to some provisory loop. The spirem is a single and continuous thread, and if endings were present they would be seen at least in those last two figures where every detail in our preparations was exceptionally clear. It is easy now to anticipate what would happen were those figures cut 5 or 8μ in thickness. The tips of the loops would be cut off, and figures 59 to 61 would give a remarkable example of many pairs of threads appearing as if they were to perform the conjugation sought by the parasyndeticians. Jules Berghs[1] stated that he found such structures from the very beginning of the prophasic movement. From our material it seems that the parallel filaments are simply the legs of the loops placed side by side, superimposed, or more or less twisted one with the other. This is probably the reason why some of them are more tardy in their movement towards the center since the crossing of the threads is a second force to be overcome.

My conclusion is, therefore, that the presynaptic spirem is single and continuous throughout and that its contraction towards synapsis gives rise to a certain number of coils and loops the heads of which are the last parts to separate from the nuclear wall. Synapsis reaches its full stage with figure 62, the most. characteristic aspect of the prophasic movement. All the chromatic substance of the nucleus, sometimes even the nucleoli, are massed in the center of the nuclear cavity as a compact knot. Only some projecting threads are to be seen in its edges. Whether this black and massed form is absolutely natural or partially the effect of the stains is difficult to tell. It is certainly the most common type of synapsis, in the most different materials and under the most varied technical conditions.

Figure 63 shows the main characteristics of the preceding figures, that is, a single thread in synaptic condition. The

[1] J. Berghs. La formation des chromosomes hétérotipyque dans la sporogénèse végétal. II. Depuis la sporogonie jusq'au spiréme définitif dans la microsporogénèse de l'*Allium fistulosum*. La Cellule, *21* (1904).

threads are, however, larger in diameter and the chromomera are bigger and more regularly distributed. In my opinion this figure represents a stage more advanced than that of figure 62 and is the starting point of the expansive movement of the spirem after full synapsis. This expansion proceeds gradually and is just the opposite of what was observed in the phases preparatory to synapsis. Figures 63 to 65 show the different steps taken by the spirem until it occupies again the whole nuclear cavity. This last figure marks the end of synapsis. The spirem, single throughout the entire phase, and with a chain of single chromatic granules, is shown in a cut condition in the last figures as a result of the larger size of the megasporocytes. In the stages previous to synapsis the spirem is very slender, the chromatic granules very small and more or less irregularly distributed (Figs. 59 to 62). From figures 63 to 65 the spirem enlarges much in diameter, the chromatin granules become bigger and distributed more regularly throughout. How did the spirem enlarge in diameter? Possibly by contraction in length, but more naturally by the absorption of food material either suspended in the enchylema in the form of diffused chromatin substance or carried from the outside through the nuclear sap. This is not a mystery. Just as the cell grows from the beginning to the end of synapsis, so the nuclear contents will grow too. And as the stage represented in figure 62 lasts for a long period, it is natural that the growth in diameter should be greater than in the other preceding periods.

*B*. From synapsis to metakinesis.—As has been stated, in the microsporocyte there is a very close gradation between the first and second contractions which enables us to trace very accurately the transition from one to the other. Unhappily this was not the case for the megasporocyte. There were many figures of the two characteristic contractions even in the same ovary, but none showing an intermediate stage. Furthermore, those representing the second contraction were of a very peculiar character. Their chromosomes showed an excessively granular condition which leaves me in doubt as to whether this condition is natural or merely a product of the reagents acting upon excessively sensitive structures. Furthermore, they are much broken, as a rule, crowded, or apparently fused in some places. Whether this shows a state of disintegration of the chromosomes due to delicacy of structure, or whether it is rather a peculiar morphological condition, it is difficult to tell. The last view seems more probable as the surrounding tissue was in perfect condition.

Figures 66 and 67 show what I think to be the full stage of second synapsis. Apparently the behaviour of the nuclear structure is this: the spirem of figure 65 suffers a second contraction and proceeds to dispose itself in the specific reduced number of chromosomes which begin to project as loops towards the periphery, until the whole spirem is used up to form twelve more or less individualized rays, which later will become the twelve definite bivalent chromosomes. After the spirem is entirely developed into a definite number of projecting loops with many twists, each loop begins to contract by itself, reducing the number of coils and at the same time separating from the others and giving rise finally to a diakinetic chromosome (Fig. 68). The granular condition of the chromosomes prevails up to the end of the second contraction but is more accentuated in its first stages, which gives an indication that these phenomena are natural.

The behaviour of the nucleolus up to the stage just mentioned is the same as that observed in the microsporocyte. From the stage represented by figure 67, however, the nucleoli again begin to appear vacuolated as before, and to divide as if by a process of self-reproduction, into new micronucleoli. This self-division continues with the forward development of the chromosomes and at the same time the number of the vacuoles begins to diminish. Sometimes they accompany the new micronucleoli (Fig. 68).

After the chromosomes separate they spread freely throughout the nuclear cavity. In figure 68 they are exactly twelve in number without any definite position or regular orientation. This stage, however, does not last long. Almost simultaneously there is a movement of orientation of the chromosomes, the rupture of the nuclear wall with a slight invasion of the nuclear cavity by the general cytoplasm, and the dividing of the nucleoli into numerous micronucleoli and their accompanying expulsion from the field assigned to the chromosomes into the general cytoplasm of the cell. In figure 69 three chromosomes have their heads turned to the center while the others are in the normal position peculiar to this stage, as was found to be the case in the microsporocyte. This feature was not found in any other figure but it must be attributed, I believe, to the fact that the mother star stage was not fully reached, the chromosomes not being yet entirely oriented.

In this same figure there is a chromosome quite different from the others. It is arranged in three loops, the whole bivalent chromosome being a little longer than the average of the others. Whether there is any special significance attached to this structure I do not know, but such variations in form are not very rare, and are met with principally in the microsporocyte.

*C.* From metakinesis to the daughter star stage.—The manner in which the chromosomes place themselves in the spindle fibres to perform their transverse division and follow their way towards the poles of the figures is exactly the same as that already stated for the microsporocyte (Fig. 70). The spindle fibres retract towards the poles, the loops untwist and straighten along the meridian of the figure (Fig. 71) and finally break apart and separate into two equivalent branches, whereby the reduction division is accomplished.

The last figure is worthy of a little more consideration inasmuch as it affords us the most striking instance of the process by which the first division takes place. What we obsérve there is a complete confirmation of the looping hypothesis of the diakinetic chromosomes and their transverse division in the first metakinesis. The large chromosome of figure 71, perpendicular to the equator and projecting toward both poles to an equal distance from the center, has not yet performed its transversal division, while its companions are already in the anaphase stage. Its length is surely twice that of its neighbors and it is being pulled upon with equal force by both groups of spindle fibres. The untwisting of this chromosome is complete and it remains only for it to be broken through the spindle for the daughter chromosomes to proceed to the poles. I must note besides, that this chromosome being in the center of the slice, was untouched by the knife nor was it in any way affected by manipulation, and that the remarkable neatness and perfection of all the minute structures of the cell assure us that the technique used was as perfect as it could be. These remarks apply equally to the chromosome of the microsporocyte represented in figure 29. This gives even more evidence of the looping theory by the fact of its thinness over the point through which it is to divide.

Between figures 71 and 72 all attempts to find an intermediate stage have failed. However, from what we have said about the microsporocyte it is easy to see what would have taken place in the megasporocyte. Figure 72 may be considered as homologous with figure 30. The daughter nuclei are already reconstructed or nearly so. The chromosomes are more or less mixed and confused with one another and joined together by means of protoplasmic fibres.

We may now turn to the nucleoli. When the nuclear wall of the megasporocyte breaks away and the diakinetic chromosomes arrange themselves in a star around the equator of the figure, the nucleoli, already broken into numberless micronucleoli, invade the cytoplasm (Fig. 69), and continue to move away as the two daughter nuclei separate and reconstitute themselves (Fig. 71). When the daughter nuclei are formed, the micronucleoli, spread throughout the cytoplasm, begin a backward movement toward the new nuclei, and begin to center within them until they become complete and ready to start the second division. On the other hand the more the new micronuclei reach the daughter nuclei and enter within them, the more they fuse together, also, growing much in size and diminishing in numbers. After the nucleoli are thus within the nuclei the aspects of the latter are somewhat the same as those of the nuclei of the megasporocyte in its first prophasic movements.

D. FROM SECOND PROPHASE TO THE COMPLETION OF THE EMBRYOSAC.—The second and third divisions of the nuclei of the embryosac follow one another in very rapid succession. The manner in which these divisions take place (Figs. 73 to 78) is the same in character as that described for the microsporocyte with the difference that three divisions are performed successively for the formation of the oosphere. Furthermore both the daughter and the grand-daughter nuclei, which result from those divisions, remain embedded in the general cytoplasm of the embryosac without separating cell walls, while in the microsporocyte each nucleus takes a part of the general cytoplasm for itself and becomes separated from its neighbors by a new secondary cell wall.

The movement of the daughter nuclei from the first telophase to the beginning of the second prophase may be summarized as follows: After the reduction division takes place (Fig. 71), the chromosomes of each daughter nucleus come together and begin to form a knot (Fig. 72) which seems to grow in volume or rather to inflate, while the chromosomes seem to mingle themselves more intimately and join one another by means of many protoplasmic connecting fibres. This condition becomes more accentuated at the end of the telophase when the nucleoli have already completely entered the daughter nuclei. This is, I think, the so-called resting stage of the daughter nuclei where a reconstitution of the nuclei is admitted by many cytologists. I can not believe in such reconstitution. According to my observations both in the microsporocyte and megasporocyte the chromosomes of both sporocytes remain independent and autonomous from the first to the last division. The apparent resting stage lasts but a short time. The prophasic movement of the second division begins very close to the end of the first telophase. This movement begins with the breaking down and expulsion of the nucleoli and at the same time the chromosomes of the daughter nuclei get free from their anastomoses (protoplasmic connecting fibres) and prepare themselves to form the new equatorial plates.

The succeeding phenomena which take place in the further development of this phase of the embryosac (Fig. 73) are essentially the same as those already observed in the microsporocyte. The chromosomes resulting from this second division become shorter and grow in diameter, and proceed towards the respective poles in a more or less straight line (Fig. 74), exactly as they were seen to do in the same stage of the microsporocyte, until they reach the telophase stage (Fig. 75).

During this second division the same steps are observed in the movements of the nucleoli as during the first mitosis. They are broken down and expelled in the beginning of the prophase and remain in the cytoplasm until the chromosomes reach the second telophase (Fig. 75). From this point to that represented in figure 76 the behaviour of the chromosomes and of the nuclei themselves, as well as that of the nucleoli, is exactly the same as that observed in the analogous phases of the first mitosis.

Figure 77 shows the last phase of the third mitosis in the embryosac. There is nothing particular to be noted in connection with this division. The same process of division of the chromosomes and the same diversity in regard to the plane of division recorded in the second mitosis are taking place. The only thing to be noted in this figure is that its lower nucleus, which will become two of the antipodals, is already separated from the general cytoplasm by a cell wall. It is the most advanced in mitotic development of all the nuclei of that embryosac and this is presumably the reason why it is already separated by a cell wall, which will envelop the others as soon as they are more advanced in development with the exception of the fusion nuclei.

Once the last division is accomplished the three lower and the three upper great-grand-daughter nuclei appear as independent cells, surrounded by a cell wall, within and at each pole of the embryosac (Fig. 78).

As we have previously seen, the growth of the megasporocyte is continuous from the first moment of its appearance and proportionately faster than the ontogenic development of its nucleus up to the beginning of the first mitosis. Now, from this stage onwards the vegetative growth is even relatively greater in regard to time but not in regard to ontogenic development. This growth is, however, very closely related to the ontogenic development as greater space is necessary with each new division of the nuclei within the embryosac. After the last division takes place the embryosac attains its largest size and is ready for fecundation.

## III. FECUNDATION PHENOMENA.

As soon as the pollen grain liberates itself from the tetrad, by the breaking of the common envelope, it grows very rapidly, acquiring its typical size and form (Fig. 44). This seems, however, to be more a vegetative than an ontogenic development, for the structure of its nucleus appears to be essentially the same both in tetrad and in free condition. In the tetrad condition of the sporocyte the pollen grain has a single wall and the nucleoli are scattered in the general cytoplasm, while in the free condition the nucleoli are for the most part condensed into one to three within the nucleus and the spore wall consists of intine and exine.

After the microspore grows almost to its largest size the nucleus divides into two (Fig. 47), giving rise to the tube nucleus, which remains in the center of the spore, and to the generative nucleus, which is separated from the general cytoplasm of the spore within a lenticular cell. The two nuclei are structurally the same, but the cytoplasm of the latter appears more granulated and hyaline, the nucleus remaining in an apparent resting condition until the spore finds a suitable medium for its further development. After it comes into contact with the stigma fluid the generative nucleus divides, giving rise to two male nuclei or gametes, none being found in process of division before dehiscence of the anther nor before the tube begins to protrude from the microspore wall (Fig. 48). No details of division of the gametophytic nuclei were found in our preparations.

The formation of the tube of the spore is apparently connected with a special activity of the generative nucleus. The exine wall of the microspore nearest to the generative nucleus breaks apart, probably because of that activity, and the intine is pushed away in a papillated form, and grows further and further into a smaller or larger tube rich in cytoplasmic substance. The generative nucleus, already divided or not into two male nuclei, passes to the tube which continues to grow more or less according to the distance to be traversed to reach the embryosac. Figure 48 shows the microspore with its tube and two male nuclei lying in it, the tube nucleus remaining yet in the center of the microspore.

After the tube reaches the embryosac, it breaks at its extremity and discharges the sperms, which move freely towards the oosphere and endosperm nucleus.

The sperms of *Lilium tenuifolium* are very large and conspicuous in all their details (Figs. 49 to 51), composed of a reticulated nucleus with a very delicate cell wall (Fig. 51). They consist of a chromatic, irregularly spirillated network lying in the periphery of the nuclear sap, with big, deeply stained granules (Fig. 52), these being more commonly visible where the filaments cross.

After the eight nuclei of the embryosac have separated and are definitely constituted (Fig. 78), some special movements and aspects of development follow preparatory to the act of fecundation. The embryosac itself still grows, and the two synergids move on, placing themselves at the entrance of the micropyle more or less opposite one another. The oosphere moves in the same direction and takes a position between the synergids, more or less in contact with them, and turned towards the micropyle as if in readiness to meet the pollen tube. The synergids seem to enter very soon into a state of disintegration perhaps by imparting their substance to the oosphere in a nutritive capacity. This disintegration may begin before fecundation, and that of the cytoplasm of the oosphere follows immediately as a preparatory act for fecundation. By disintegration I mean the resolution and absorption of a part at least of the structure concerned.

As to the antipodal cells, they move on, too, towards the opposite end of the embryosac, where they place themselves in more or less close contact with one another, sometimes in a perpendicular line, but ordinarily without any regard to a definite plan. Meanwhile they begin to disintegrate, first the cytoplasm and finally the nuclei exactly as in the case of the synergids, their contents being mixed and diffused into the general cytoplasm of the embryosac (Fig. 80), or possibly to the chalaza to help its growth and development. This last view was suggested by the fact that the lower antipodal cells are those which begin to disintegrate first, their nuclei becoming deformed and dark even before the last division is accomplished. The upper nucleus of the antipodals remains ordinarily in an apparent good condition, for a greater length of time.

The fusion of the polar cells may be simultaneous with that of the sperm but it may also follow or, more commonly, precede it. In some preparations the two polar nuclei were entirely fused, so as to appear like one nucleus only, long before the sperms entered the embryosac. The fusion begins by a kind of disintegration and absorption of their adjacent walls, followed by the fusion and mixing up of their nuclear contents.

All the nuclei of the embryosac, during the phases just described, have practically an identical structure. The filaments of the network are slender and crowded in all directions, so that the nuclei resemble those of the somatic cells in their resting stage. Three to seven large nucleoli were present in the polar nuclei, the others having, as a rule, only three to five.

While these phenomena are being realized in the embryosac the pollen tube is making its way through the cells which surround the micropyle (Fig. 79). These cells are more or less closely attached to one another, so that the pollen tube must force its way through them, becoming more or less undulating in shape (Figs. 79 and 80). Once arrived at the entrance of the embryosac it passes sometimes through the synergids or remains more or less in contact with them. Meanwhile the lower extremity of the tube disintegrates and the sperms are discharged into the embryosac where they continue to move freely each one toward the nucleus which is to be fecundated.[1]

The phenomenon of fecundation is exclusively performed between male and female nuclei, the general cytoplasm having no direct part in it, unless as a nutritive or conveying medium. The sperms are slightly different in size. The smaller one fuses with the oosphere while the other goes to join the endosperm nucleus. Whether this be of real significance or not I do not know, but the mere fact that the later sperm must traverse a longer way than the former to reach its proper nucleus, and assuming that during such movement it assimilates food material from the general cytoplasm, it is but natural that it should grow more in size than the former. In either case, however, its

---

[1] After the sperms are discharged into the embryosac there remain either in the tip of the pollen-tube or near it in the cytoplasm of the embryosac two dark bodies irregular in shape. What significance they have I am at present unable to state.

structure is absolutely the same and there is nothing by which an essential distinction may be drawn between the two male gametes.

The fusion of the sperms in the case of both oosphere and endosperm nucleus, is slow and gradual and seems to proceed in the same way as that already noted for the two polar nuclei.

# CONCLUSIONS.

1. The spirem comes directly from the nuclear network of the cell resulting from the last sporophytic division.

2. The spirem is a single and continuous filament with a single chain of chromatin granules distributed throughout with more or less regularity.

3. The twenty-four somatic chromosomes of the last sporophytic division fuse end to end to form a single and continuous spirem.

4. After the spirem is fully developed it begins to separate into the form of indefinite loops from the nuclear wall and concentrates at some place within the nuclear cavity; it remains for a while in this condition, growing simultaneously in diameter, and expands again in the form of loops throughout the whole nuclear cavity.

5. These movements, which constitute the synaptic stage, consist of three different phases: contraction of the filaments, apparent rest in which they grow in diameter, and expansion.

6. The thick spirem proceeds from the thin by a gradual increase in diameter chiefly during synapsis and never by a conjugation of any two spirematic filaments.

7. Synapsis is most probably a normal stage of the maturation mitosis and not the effect of reagents and manipulation. Synapsis was seen in living cells.

8. The function of synapsis is probably to bring the spirem into contact with the chromatic substance diffused in the enchylema for a nutritive purpose.

9. After synapsis a second contraction takes place which seems to play an important part in the disposition of the spirem into twelve loops before the chromosomes separate from one another.

10. Between the two said contractions an actual cleavage of the spirem was observed both in the microsporocyte and megasporocyte. The instances, however, were very few and therefore not very conclusive.

11. The longitudinal cleavage of the thick spirem is the

manifestation of the somatic tendency of the component elements of the simple spirem.

12. During the second contraction the spirem orients itself into twelve loops radiating from a closely entangled central mass.

13. After synapsis and during the whole period of the second contraction the chromosomes are being formed by the longitudinal contraction and twisting of the loops that are to give rise to them.

14. The twelve loops break apart into twelve chromosomes which spread throughout the nuclear cavity during the diakinetic stage.

15. The definite bivalent chromosomes divide transversely during the metaphase.

16. Each definite bivalent chromosome becomes attached at the extremities of its legs to the spindle fibres and breaks at the middle into two equivalent chromosomes, each of which moves to an opposite pole.

17. Each loop or diakinetic chromosome represents two somatic chromosomes fused end to end, and the first mitosis is therefore a reduction division.

18. During telophase (first and second) there is no true reconstitution of the nuclei, but the chromosomes remain autonomous from the first to second and third divisions as the case may be.

19. The division of the chromosomes in the second mitosis (and third for the megasporocyte) is longitudinal or typical, and therefore equational.

20. The nucleolus appears invariably in every nucleus and accompanies the evolution of the same by an evolutive cycle, especially its own, which cycle is repeated in every mitosis.

NOV 1 2 1915

# EXPLANATION OF THE FIGURES.

Plates I to III were reduced to half, and IV to VI to one-third in reproduction. The figures were studied with Bausch and Lomb and Leitz microscopes. Oil immersion objectives were employed throughout with a few exceptions to be noted in their proper places. The numbers of the oculars and objectives will be given after the figures, it being understood that those of Leitz were respectively compensating and apochromatic. In the same way the microscopes used will be marked by the initials B. & L., and L. respectively. The figures were drawn with a camera lucida. In a few instances, where Heidenhain's iron-alum-haemotoxylin was employed, special mention will be made after the description of the figures.

PLATE I.

Fig. 1. Microsporocyte with chromatin network beginning to form the spirem.   L. 8 x 2 mm.

Fig. 2. More advanced stage of the same showing prominent chromatin granules on the delicate threads.   L. 8 x 2 mm.

Fig. 3. Still more advanced stage showing threads being freed from their anastomoses.   L. 8 x 2 mm.

Fig. 4. Much advanced stage of the microsporocyte showing threads quite free from their anastomoses and beginning to separate from the nuclear wall.   Iron-alum-haem.   L. 8 x 2 mm.

Figs. 5–6. Microsporocytes just beginning to separate with looped continuous spirem in its movement towards synapsis. L. 8 x 2 mm.

Fig. 7. Some loops of that stage.   L. 12 x 2 mm.

Fig. 8. Full synaptic stage with the spirem entirely contracted into a central knot.   B. & L. 10 x 1.9 mm.

Fig. 9. Synaptic stage examined and drawn from living material.   B. & L. 15 x 4 mm.

Fig. 10. The beginning of the movement of expansion of the threads after the nucleus reached the full synaptic stage.   L. 8 x 2 mm.

Figs. 11–12. Successive stages in the spirem in its movement of expansion after full synapsis.   L. 8 x 2 mm.

Fig. 13. Microsporocyte going out of synapsis showing threads of three different thicknesses.   L. 8 x 2 mm.

Fig. 14. Some details of the same.   L. 12 x 2 mm.

Fig. 15. Microsporocyte immediately after synapsis with prominent spirem and single row of chromatin granules.   L. 8 x 2 mm.

Fig. 16. Microsporocyte, some time after synapsis, beginning the second contraction.   The spirem is arranging itself into twelve loops which will become the definite chromosomes. L. 8 x 2 mm.

PLATE I.

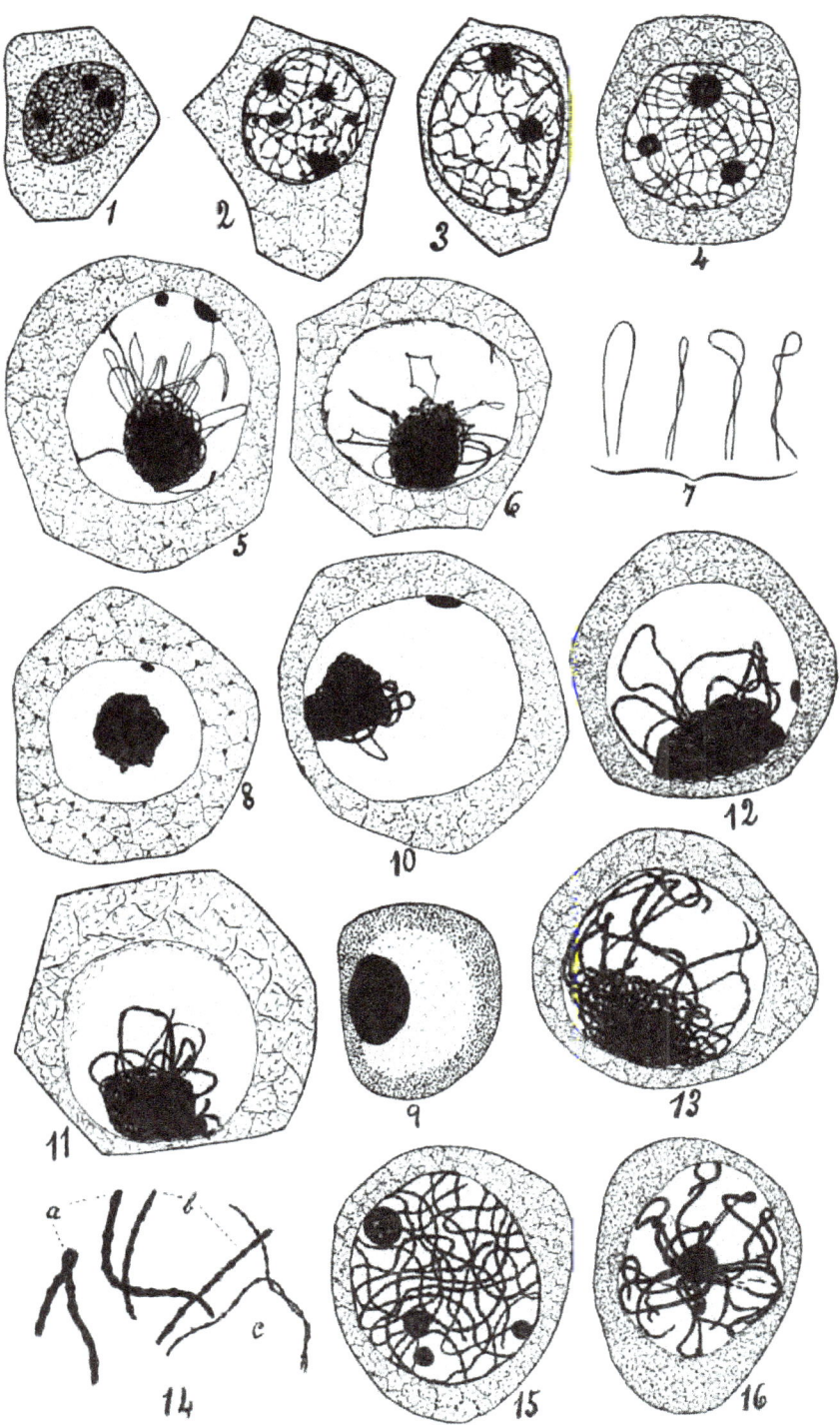

D. DA CRUZ del.

*Lilium tenuifolium.*

PLATE II.

Figs. 17–18. More advanced stages of the second contraction where the chromosomes are separating.  L. 8 x 2 mm.

Fig. 19. Microsporocyte in the stage of diakinetic chromosomes.  L. 8 x 2 mm.

Fig. 20. Microsporocyte in the equatorial plate stage with twelve chromosomes.  L. 8 x 2 mm.

Fig. 21. Mother star showing position of the chromosomes. L. 8 x 2 mm.

Figs. 22–25. Individual chromosomes of the same stage showing the process and some phases of their untwisting and unlooping. L. 12 x 2 mm.

Fig. 26. Daughter star stage.  L. 8 x 2 mm.

Fig. 27. Microsporocyte reaching the end of anaphase. L. 8 x 2. mm.

Fig. 28. Late anaphase showing the beginning of the formation of the mid-body at the equator of the spindle.  L. 8 x 2 mm.

Fig. 29. Loose daughter skein stage with an entire chromosome placed across the equator in process of transverse division. L. 8 x 2 mm.

Fig. 30. Loose daughter skein stage.  L. 8 x 2 mm.

Fig. 31. Daughter nuclei completely separated and beginning second division.  L. 8 x 2 mm.

Fig. 32. Daughter star of the second division with the chromosomes beginning to separate.  L. 8 x 2 mm.

Figs. 33–38. Individual chromosomes showing the process of division in the preceding stage.  L. 12 x 2 mm.

Fig. 39. Second anaphase where the chromosomes are separating.  L. 8 x 2 mm.

PLATE II.

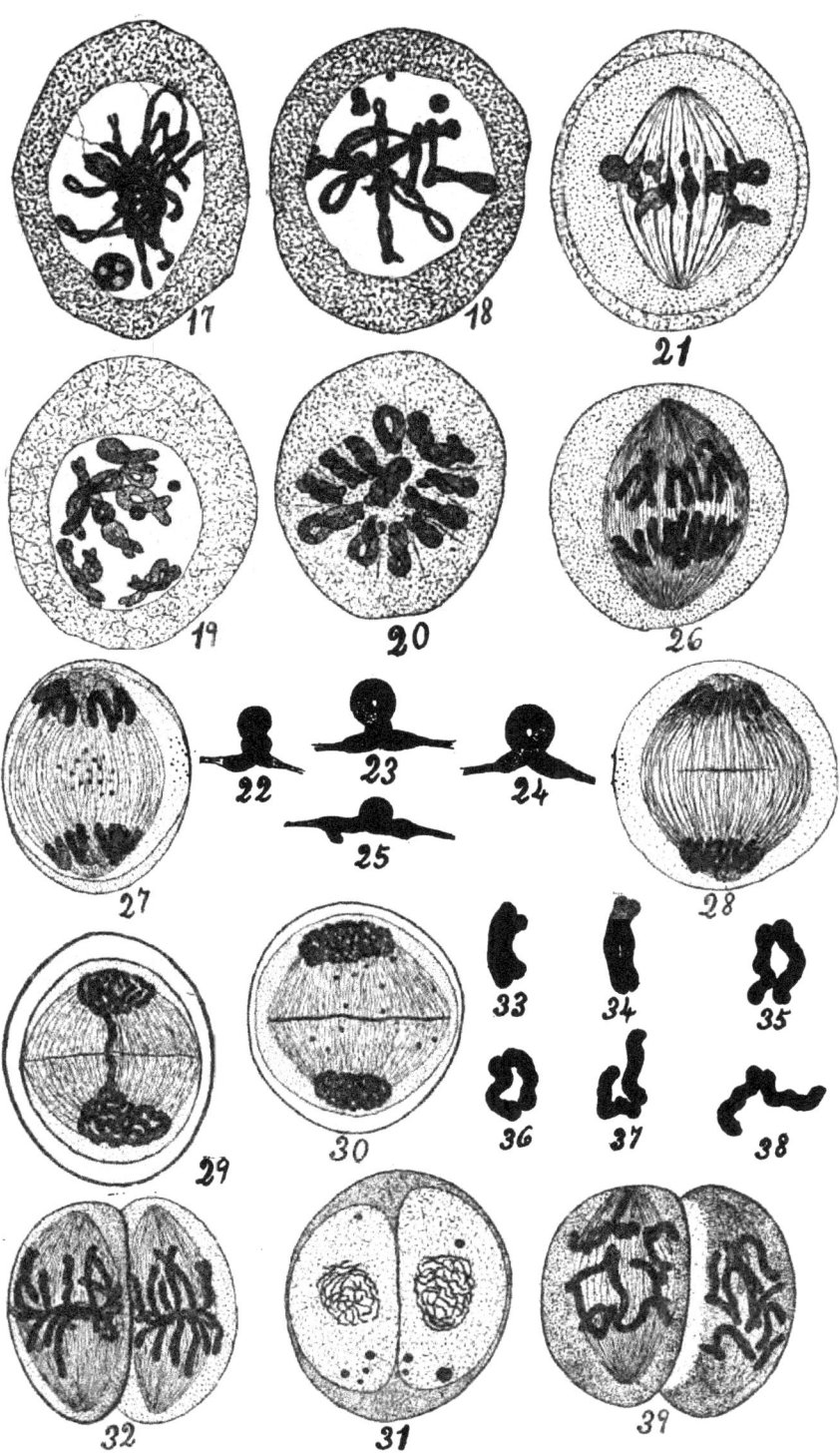

D. DA CRUZ del.

*Lilium tenuifolium.*

## PLATE III.

Fig. 40. More advanced stage of the second division. Iron-alum-haem. L. 8 x 2 mm.

Fig. 41. Late anaphase of the second division showing two chromosomes just separating from one another. L. 8 x 2 mm.

Fig. 42. Loose grand-daughter skein with micronucleoli throughout the cytoplasm and spindle. Iron-alum-haem. L. 8 x 2 mm.

Fig. 43. Tetrad at the end of second division with the nucleoli in the cytoplasm. L. 8 x 2 mm.

Fig. 44. Whole single pollen grain. L. 8 x 2 mm.

Fig. 45. Longitudinal section of the microspore just about to germinate, showing nucleolus in the cytoplasm and other in the nucleus. L. 8 x 2 mm.

Fig. 46. Microspore in cross section. L. 8 x 2 mm.

Fig. 47. Germinated microspore showing generative and tube nuclei separated by a lenticular wall. L. 8 x 2 mm.

Fig. 48. Pollen grain with developed pollen tube containing tube nucleus in the center and two sperms descending the tube, these not very distinct in their details. B. & L. 10 x 1.9 mm.

Figs. 49–50. Whole sperms after discharged into the embryo-sac. 50: L. 12 x 2 mm. 49: B. & L. 10 x 1.9.

Fig. 51. Superficial view of a sperm in cross section. L. 12 x 2 mm.

Fig. 52. Two filaments of the reticulum of a sperm. B. & L. 15 x 1.9.

PLATE III.

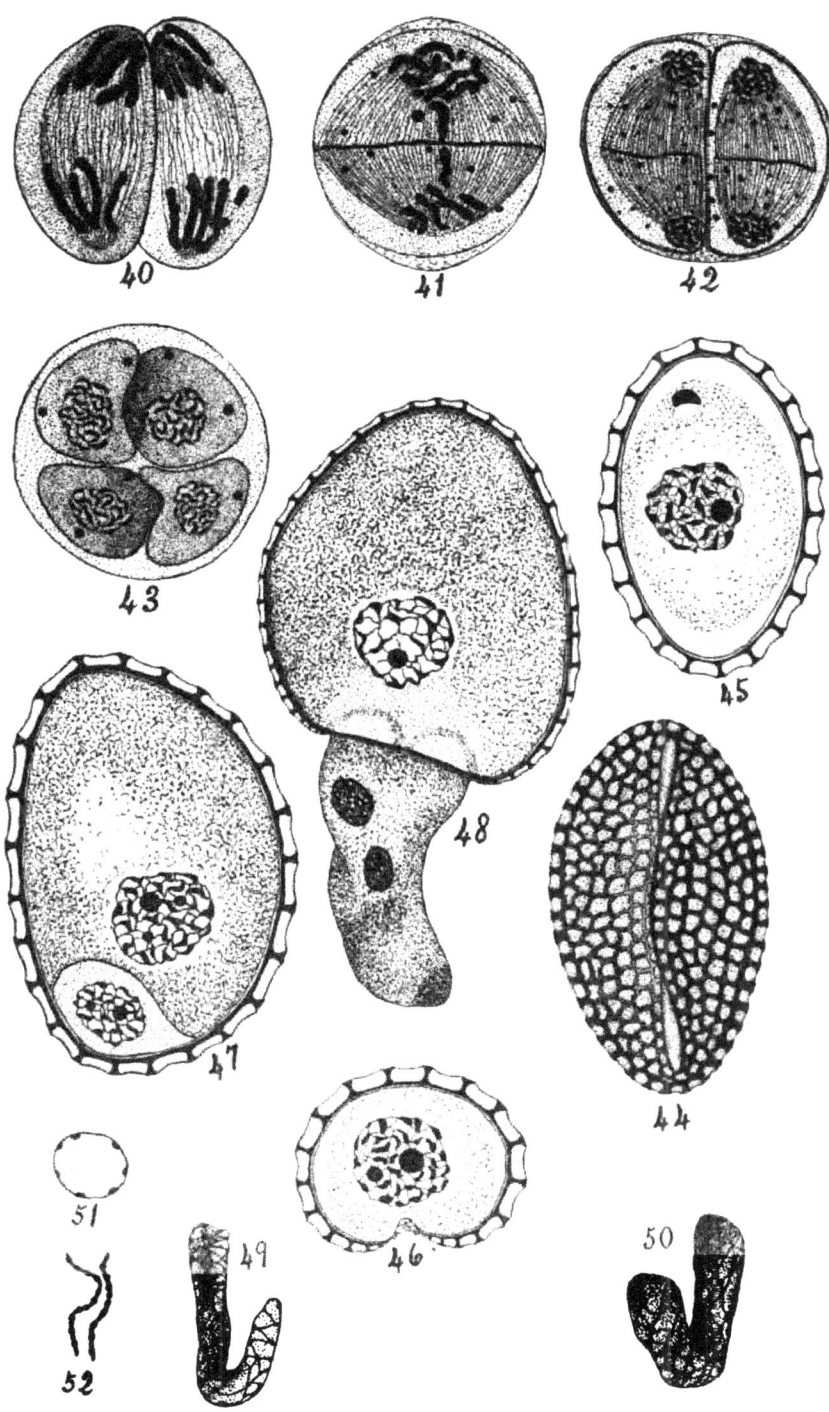

*Lilium tenuifolium.*

# PLATE IV.

Fig. 53. Megasporocyte showing chromatin network beginning to form the spirem.   L. 8 x 2 mm.

Figs. 54–55. Successive stages of the megasporocyte showing chromatin granules on the delicate threads.   B. & L. 10 x 1.9 mm.

Fig. 56.   Practically the same stage stained in iron-alum-haematoxylin.   L. 8 x 2 mm.

Fig. 57. Older megasporocyte with chromatin threads showing about the same development as Fig. 56 and with much-vacuolated nucleoli.   L. 8 x 2 mm.

Fig. 58. More advanced stage showing chromatin threads just beginning to contract in form of loops.   L. 8 x 2 mm.

Fig. 59. Megasporocyte with chromatin threads separating from the nuclear wall to be thrown into synapsis.   L. 8 x 2 mm.

Fig. 60. More advanced stage than before with some loops already separated from the nuclear wall.   L. 8 x 2 mm.

Fig. 61. Megasporocyte with chromatin threads separated from the nuclear wall and moving towards the center in form of loops to reach synaptic stage.   L. 8 x 2 mm.

Fig. 62. Megasporocyte reaching full synapsis. L. 8 x 2 mm.

Fig. 63. Synaptic stage where the spirem, in form of many entangled loops, larger in diameter than in the previous stages, begin to extend again by a gradual and characteristic movement of expansion.   L. 8 x 2 mm.

Fig. 64. Megasporocyte showing looped spirem in its movement of expansion after the full synaptic stage.   L. 8 x 2 mm.

Fig. 65. End of synapsis in which the expansive movement of the thick spirem ceases.   L. 8 x 2 mm.

PLATE IV.

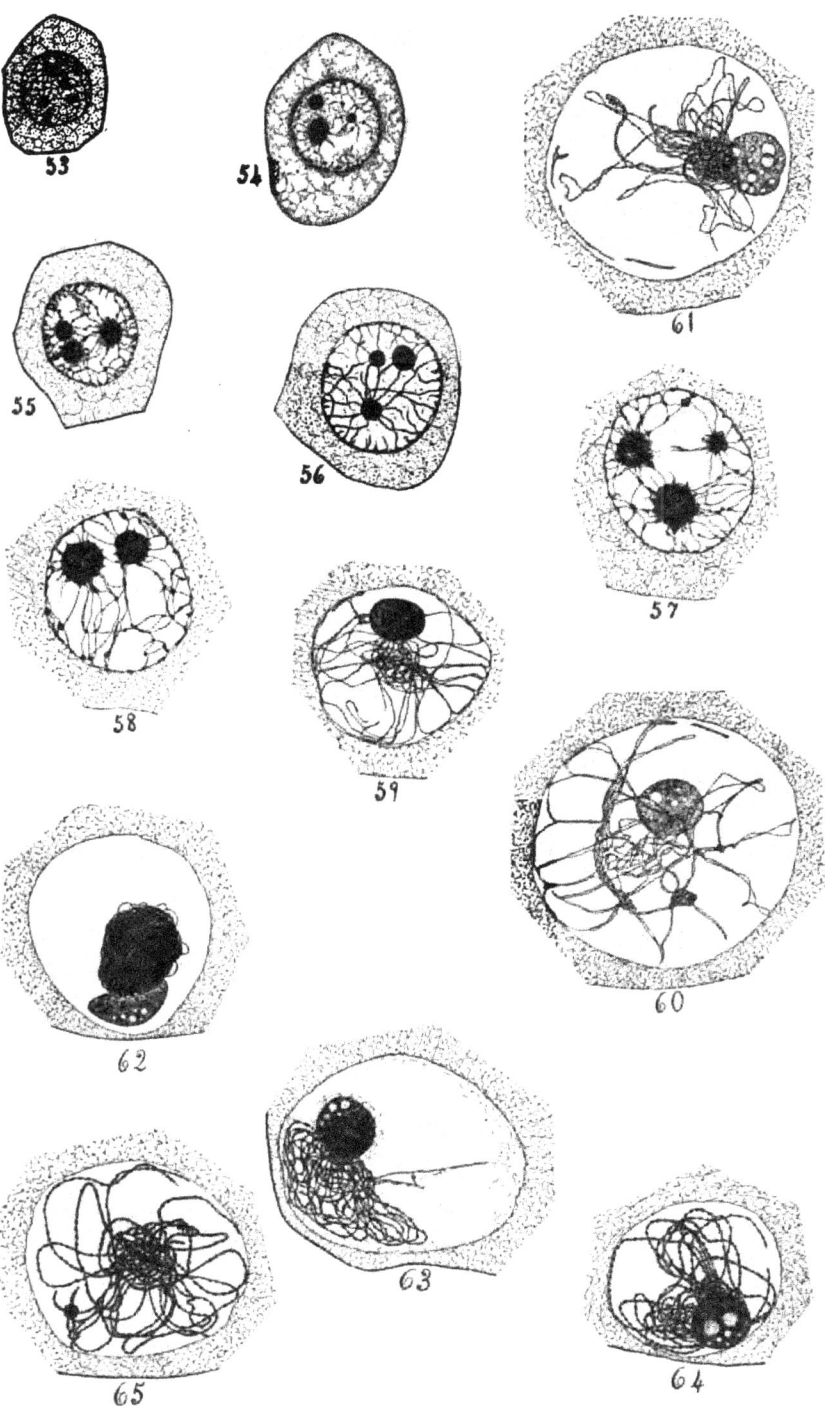

*Lilium tenuifolium.*

## PLATE V.

Figs. 66–67. Two stages of the second contraction in which the chromosomes are exceedingly granular in structure and the nucleoli uniformly stained.  L. 8 x 2 mm.

Fig. 68. Megasporocyte with twelve diakinetic chromosomes irregularly distributed throughout the whole nuclear cavity. L. 8 x 2 mm.

Fig. 69. Megasporocyte showing twelve looped chromosomes forming the equatorial plate.  The nucleoli were broken into many micronucleoli and scattered throughout the cytoplasm. L. 8 x 2 mm.

Fig. 70. Practically the same stage in longitudinal section. B. & L. 10 x 1.9 mm.

Figs. 71–71a. The only two slices of the same megasporocyte in which all chromosomes but one are in the anaphasic stage. L. 8 x 2 mm.

Fig. 72. Daughter nuclei completely separated with micronucleoli scattered throughout the cytoplasm.  B. & L. 10 x 1.9 mm.

Fig. 73. Metaphase of the second division.   L. 8 x 2 mm.

PLATE V.

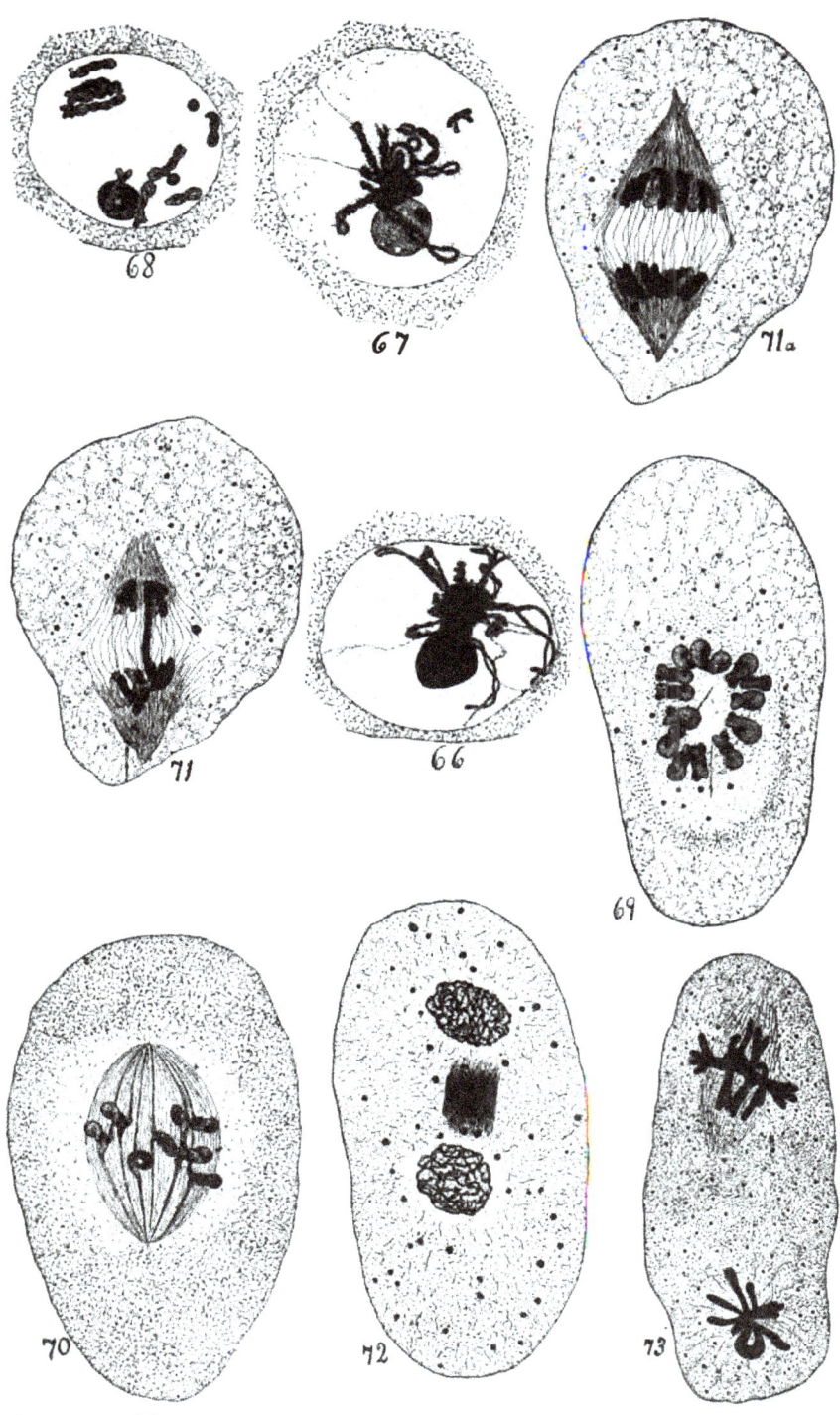

D. DA CRUZ del.

*Lilium tenuifolium.*

PLATE VI.

Fig. 74. End of the second anaphase. B. & L. 10 x 1.9 mm.

Fig. 75. Grand-daughter nuclei with vestiges of the spindle of both divisions. B. & L. 10 x 1.9 mm.

Fig. 76. More advanced stage in which the micronucleoli had entered the nuclei and fused into a limited number of nucleoli. B. & L. 10 x 1.9 mm.

Fig. 77. Embryosac showing different stages of the last division. L. 12 x 4 mm.

Fig. 78. Embryosac with eight nuclei some time after they have separated. L. 12 x 4 mm.

Fig. 79. Pollen-tube traversing the micropyle. B. & L. 5 x 1.9 mm.

Fig. 80. Embryosac with pollen-tube showing advanced triple fusion. B. & L. 10 x 4 mm.

PLATE VI.

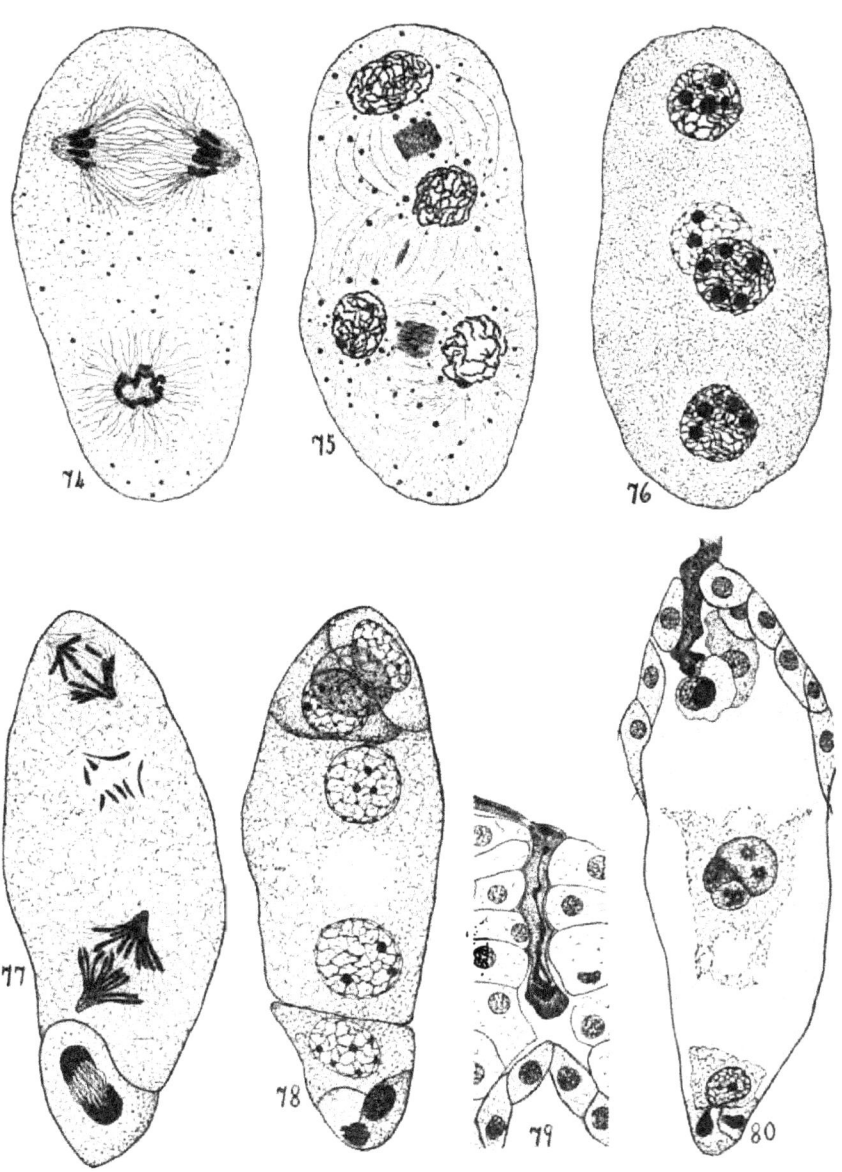

D. DA CRUZ del.

*Lilium tenuifolium.*

PLATE VII.

*a*. Photograph of figure 71, showing the whole chromosome and its chromatin granules.

*b*. Same stage as represented in figure 70, showing a good twisted loop.

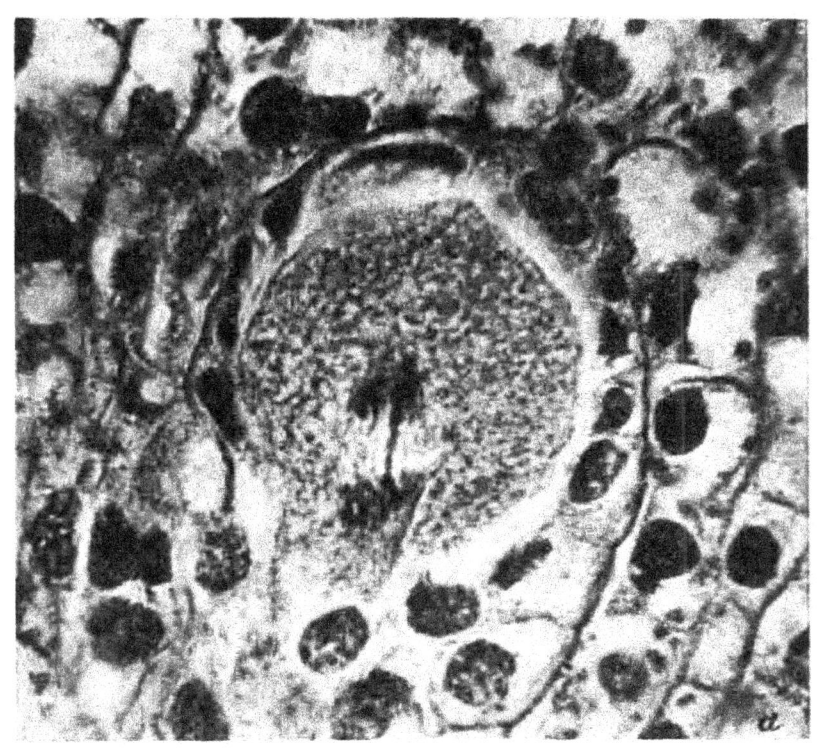

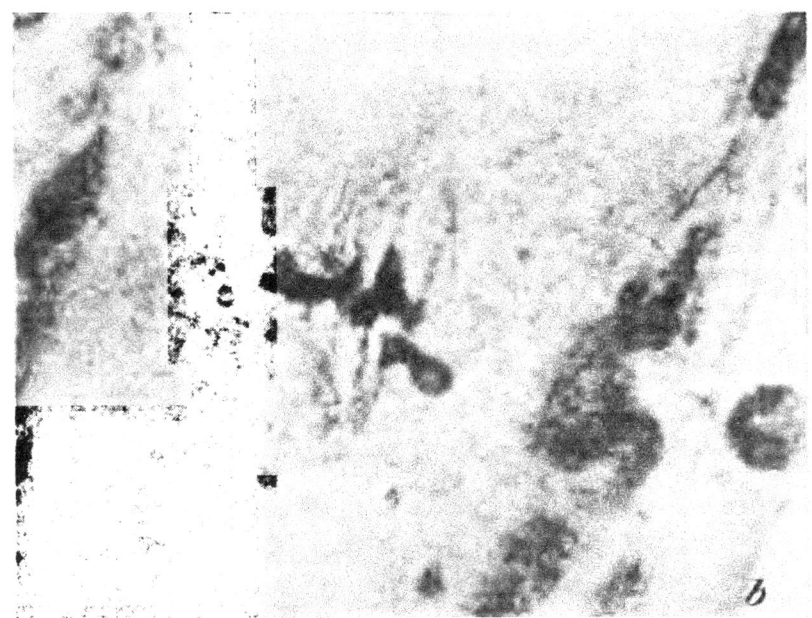

*Lilium tenuifolium.*

# VITA

Daniel da Cruz was born at Villar, District of Lisbon, Portugal, March 1st, 1880. He received his early education in the public-school course of Villar and in the two years' preparatory course at St. Bernardino College. He joined the Franciscan Order, September 1896, at Varatojo, and pursued courses of Philosophy and Sciences at St. Bernardino and Montariol, Braga, Colleges, being graduated at Montariol in 1901.

He began his course of Theology in Sevilla, Spain, the same year, and was graduated in 1905 at Braga Theological Seminary, whither he had returned in 1903. He was ordained to the priesthood in July, 1905, sent to the Mozambique, Portuguese East Africa, Missions, October, 1906, and was appointed Professor of Sciences in the Franciscan College of Leiria in 1910. He came to the United States, entered the Catholic University of America October, 1911, and pursued the courses of study in Biology, Botany, Chemistry, Philosophy and German under the following professors and instructors: Dr. J. J. Griffin, Chemistry; Dr. T. V. Moore, Philosophy; Dr. P. Gleis, German; Prof. J. B. Parker and Mr. G. J. Brilmeyer, Biology, to whom he expresses his appreciation for their sympathetic guidance of his studies.

CPSIA information can be obtained
at www.ICGtesting.com
Printed in the USA
BVHW081133210119
538274BV00023B/1005/P